D1252243

# The ABCs of Aphasia

## *A Stroke Primer*

**Thomas G. Broussard, Jr., Ph.D.**

*Johnny Appleseed of Aphasia Awareness*

*Thomas G. Broussard, Jr.*

Stroke Educator, Inc.
St. Augustine, Florida

Book and cover design by Thomas Broussard and Sagaponack Books & Design

ISBN: 978-1-7344142-0-2 (softcover)
ISBN: 978-1-7344142-2-6 (hardcover)
ISBN: 978-1-7344142-1-9 (e-book)

Library of Congress Catalog Number: 2020900715

Summary: *The ABCs of Aphasia: A Stroke Primer* includes images and aphasia-friendly definitions using the 26 letters in the English language alphabet, over 90 glossary entries, plus aphasia information such as aphasia car magnets, aphasia ID cards, aphasia caregiver info, aphasia research info, aphasia organizations' websites, and recovery plan tips, all regarding stroke, aphasia, and neuroplasticity, for educating the wider public about this little-known language disorder.

MED057000-MEDICAL / Neuroscience
SCI089000-SCIENCE / Life Sciences/Neuroscience


www.StrokeEducator.com

Stroke Educator, Inc.
St. Augustine, Florida

*To Laura,*

*Josiane, David and Will,*

*Mira, Kai, Maddie, and Jack*

"I can see it, but I can't say it!"

—a person with aphasia

# Introduction

Your loved one has had a stroke … and now has aphasia. Where do you start?

This book is written *about*, *by*, and *for* people with aphasia. This aphasia-friendly primer should be in the hands of *every* person who has just started the family journey of stroke, aphasia, and recovery.

I had my first stroke in 2011. When I lost my language, I was an associate dean at a college outside of Boston. The effects of my aphasia were that I could not read, write, or speak well. It took years for me to regain my language, and I'm here to tell you it can be done!

Most people know something about strokes, but very few know much about *aphasia*. Approximately 25% to 40% of people who have a stroke acquire aphasia. It is a communication disorder, typically from a stroke, with language problems including reading, writing, and speaking.

There are about 2.5 million people with aphasia in the United States, more than many other common conditions including cerebral palsy, multiple sclerosis, Parkinson's disease, or muscular dystrophy.

*Aphasia is a serious language disorder that few people have ever heard of!*

This book is dedicated to helping people with aphasia, their family, and caregivers, as well as educating the wider public about aphasia, as part of the "*Aim High for Aphasia!*" Aphasia Awareness Campaign.

Tom Broussard
*Johnny Appleseed of Aphasia Awareness*

# A...Aphasia

Aphasia affects your ability to read, write, and understand language, both verbal and written. About 25% to 40% of people with stroke acquire aphasia. Although it affects your language, your intellect is still the same. Age and severity affect recovery; however, regaining your language is still quite possible with motivation and practice.

**Difficulty Understanding**

**Difficulty Reading**

**Difficulty Speaking**

**Difficulty Writing**

**APHASIA**

(from Greek : without speech)

# B... Broca's area

Broca's area is a region of the brain close to the left temple that is linked to speech production and language processing. It is often called the "language center," holding the rules of grammar and syntax. A stroke in the Broca's area often causes Broca's aphasia (or expressive aphasia), the difficulty of finding words that can be "seen" in the mind but can't be expressed. (Wernicke's area is presented as well.)

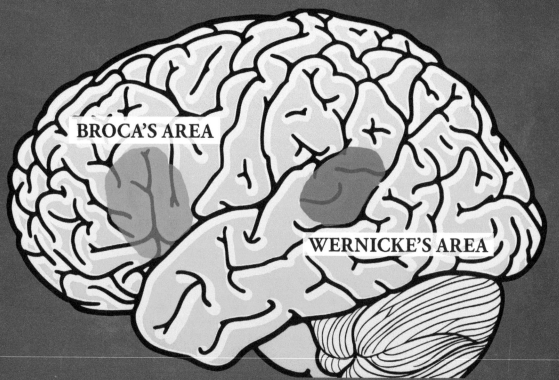

BROCA'S AREA

WERNICKE'S AREA

# C...Clot buster

Clot buster drugs (e.g., tissue plasminogen activator, tPA) are used to break up blood clots. Ischemic stroke is one of the main conditions that clot busters are used for. Typically, the drug has to be administered to someone within 3 (three) hours after a stroke, in order to restore blood flow and prevent long-term damage.

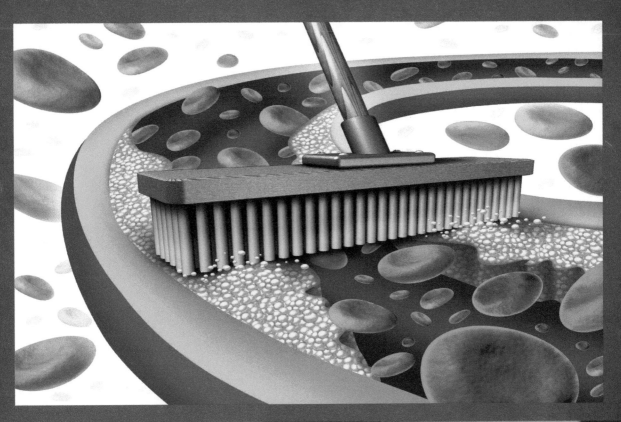

# ...Awareness of our Deficits

People with aphasia may be unaware they have the deficits that they have. Reading, writing, and speaking are skills that need to be practiced in order to get better. But if you don't know what you don't know—awareness has its own share of deficits—it keeps the other deficits from being worked on and improved. Repetitive language activities will improve awareness, as well.

# E...Exercise

Exercise, especially walking, uses a large amount of certain brain functions at the cell level (proteins and enzymes) that contribute to plasticity. As the brain continues to grow and change from doing these activities, language deficits lessen as a result. Consistent walking paired with regular cognitive activities (reading, writing, and speaking) are a boost.

# F...F.A.S.T.

It has been said that "Time is brain." The longer the brain goes without blood and oxygen, the worse it gets. It is incredibly important that the wider public become educated about stroke symptoms, what they look like, and to call for help as fast as possible, hopefully well within three hours.

FACIAL WEAKNESS

ARM WEAKNESS

SPEECH DIFFICULTY

911

TIME LOSS IS BRAIN LOSS

# G...Grammar

Our language has a lot to do with the rules that govern the structure and organization of words and sentences. People possess an internalized set of rules absorbed at childhood by hearing others in the family and friends speak. Recovery from aphasia requires more language effort and activities to "reset" and repair the cognitive structure that underlies language.

# H...Hemorrhagic stroke

This is a "bleeding" stroke (or red stroke), when bleeding occurs within the brain or outside the brain, but still within the skull. When an artery bursts, it floods the brain with blood that builds pressure within the skull, which causes damage. About 15% of stroke survivors have had this kind of stroke.

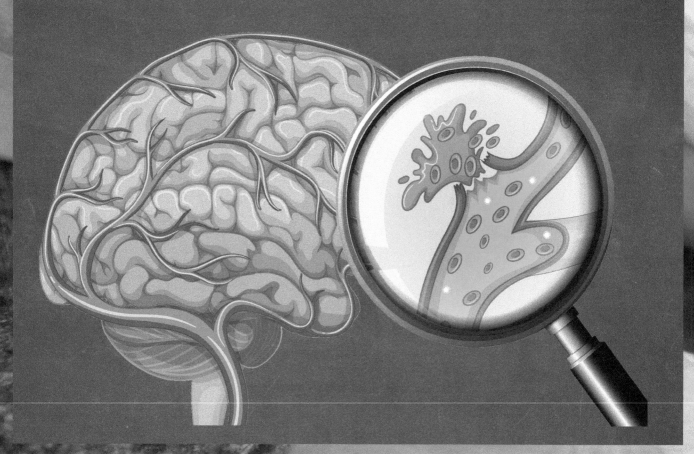

# I ...Ischemic stroke

These strokes obstruct blood flow to the brain, usually from a blood clot, and occur locally from fatty buildup (plaque) or a clot that travels from somewhere else in the body (i.e., heart, lungs). About 85% of stroke survivors have had this kind of stroke.

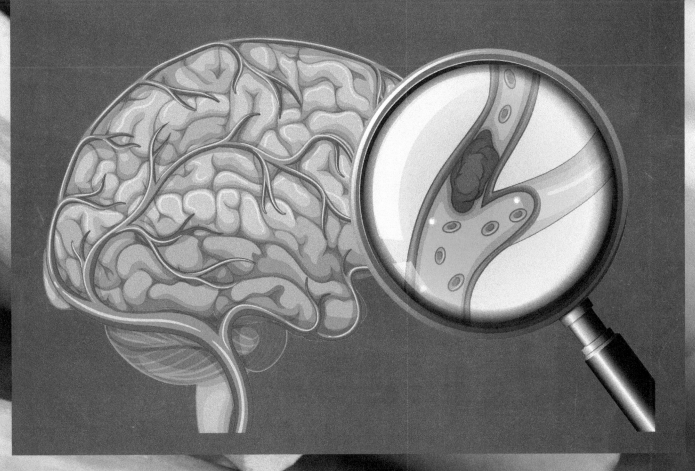

# J...Journal

Writing in a journal (or diary) is another part of regular and persistent language activities (reading, writing, and speaking) that are highly therapeutic. The activities themselves provide the needed and appropriate ingredients that establish the recipe for brain changes, as the basis for language recovery and learning.

# K...Keeping track

The brain is built to be well organized. It constructs all the routines, habits, customs, and traditions of everyday life and creates a mirrored neural copy. Once it is established, we can use calendars, schedules, logbooks, charts, diaries, and other devices that "keep track" of our deficits (among other things), and we update the ongoing neural copies as our language improves.

*Record & Listen to your voice*

*Write in your diary*

*Read aloud*

*Go for a walk*

*Take pictures*

# L ...Learning

Recovery from a stroke is a lifelong endeavor of learning. When it comes to stroke and aphasia recovery, you cannot *not* learn. Any language activities (reading, writing, and speaking), whether damaged or not, use the same learning process. Lifelong aphasia therapy is the same as personal lifelong learning, with similar results, more learning in either case.

 **M**...Motivation

The Five Rules of Aphasia Recovery
1. Motivation
2. Practice
3. Practice
4. Practice
5. Practice

# N...Neurons

These beautiful little creatures are incredibly powerful. The remaining neurons (nerve cells) after a stroke have the capacity (plasticity) to grow new, additional dendrites (branches) and synapses (leaves) to reconnect the messages that the stroke has destroyed. Once rebuilt and connected, language starts to flow again.

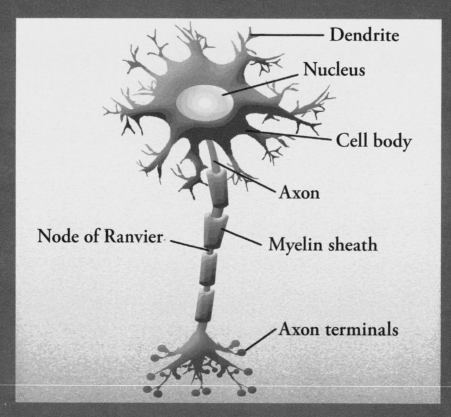

Dendrite

Nucleus

Cell body

Axon

Node of Ranvier

Myelin sheath

Axon terminals

# O...Occupational therapy

This therapy focuses on helping people return to meaningful activities and work after a stroke or other disability; that includes reviewing a person's environment and day-to-day living activities. The purpose is to regain as much function and independence as possible, whether it be driving, working, going on vacation, paying bills, or talking with family and friends.

# P...Plasticity

The brain has the capacity to change (which is to say, learn and remember) continuously for all people, whether healthy or not. Aphasia recovery is largely dependent on severity and the amount of cognitive stimuli of various language activities, including reading, writing, and speech. Plasticity is the foundation of all learning and improvement.

# Q...Quiet

For people with aphasia, background noise has a big impact on communicating. It isn't because they can't hear. The problem is that the capacity to identify and "sort out" various sounds is damaged, so that the timing and sequence of numbers, letters, words, and sentences give the appearance of not being able to hear, when it has nothing to do with one's auditory hearing.

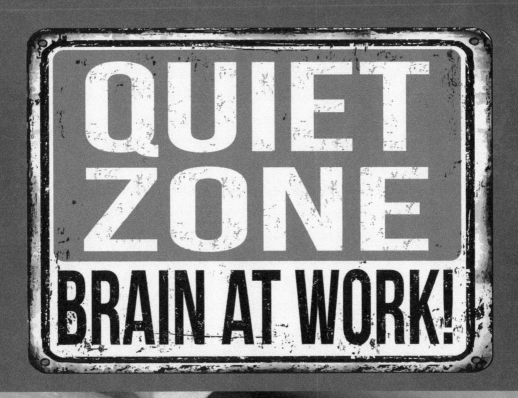

# R...Reading

One part of the problem of reading is the damaged timing that prevents one from processing all the steps leading to a full sentence. The first step is realizing that you can likely see, read, and understand individual words on their own. Reading one or two words is a start. Reading more words leads to increased transmission speed and, ultimately, reading improvement.

# S...Speech therapists

Also known as speech-language pathologists (SLPs), assess, diagnose, and treat people who have a variety of diseases and disorders (stroke, brain injury, hearing loss, and others) that include aphasia and difficulties with reading, writing, speaking, plus swallowing impairments. Speech therapists are partners with people with aphasia at the start of a lifelong journey of aphasia recovery.

# T ...Telegraphic speech

This symptom is common in nonfluent aphasia (Broca's aphasia) which is caused by a stroke. Content words (nouns, verbs) are used, but sentences are difficult to express. The person uses "telegraphic speech" with content words and without conjunctions (e.g., and, but, if) and articles (the, a, an).

WANT DRIVE STORE

# U ...Unexplained (cryptogenic) stroke

An ischemic stroke is caused by a blood clot that blocks the flow of blood to the brain. In some instances, the cause of a stroke cannot be determined. In that case, the stroke is of an unknown origin and is called a "cryptogenic (unexplained) stroke." It's estimated that about 25% to 30% of ischemic strokes are cryptogenic.

# Sometimes there is just no explanation

# V...Voice recording

This is one more tool in aphasia recovery. Recording the voice of a person with aphasia provides instantaneous feedback of one's spoken word. The taped evidence provides a feedback loop that allows the person with speech deficits to hear the deficits (and become more aware of the errors) in real time.

# W...Writing

Writing is an interesting problem for people having trouble forming letters, words, or sentences. Many people might also be unaware of the errors they are producing. However, writing anything in a diary or journal (even if you cannot write properly) induces neural feedback below the conscious level, encouraging the remaining cells to go about the business of fixing the damage on their own.

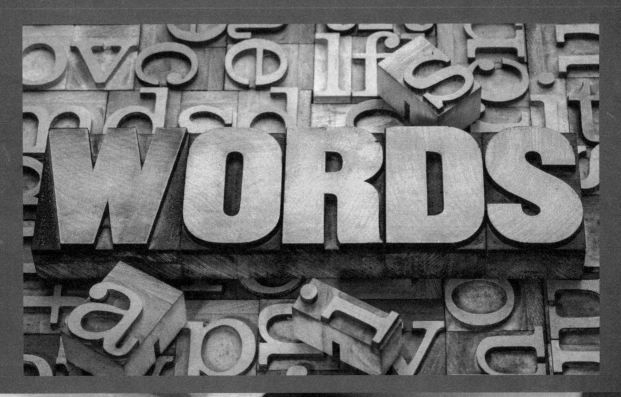

# X...EXpressive aphasia

Expressive aphasia is described as the partial loss of language (spoken or written), although comprehension is generally intact. A person with these symptoms can display difficult, effortful speech. It includes meaningful content words, but the connective words such as prepositions (at, on, to) and articles (the, a, and an) often aren't produced.

I can't find that word!

# Y... PrimarY progressive aphasia

This is a neurological condition that causes an individual's language capabilities to become progressively more impaired over time. Primary progressive aphasia is not caused by brain injury or stroke, but is a neurodegenerative disease. The condition occurs when the brain cells related to language and speech (often in the Broca's area) deteriorate.

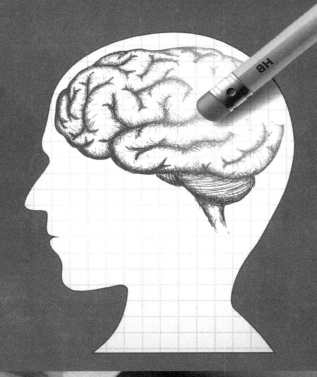

# Z...Zillions

The brain is an amazingly complex device consisting of untold numbers—such as Carl Sagan's "billions and billions"—of connections still being constantly built. It is organized in such a way that every activity on the outside induces and creates neural (brain) representations (essentially, copies) on the inside. The brain is built to learn … or repair what was built before. This is how people with aphasia get better, with plasticity ink.™

## Quantity is a quality all its own

50 to 10,000 neurons per neuronal group
+1,000,000 neuronal groups
100,000,000,000 neurons (100 billion)
100 to 10,000 synapses per neuron
+100 different neurotransmitters per synapse
4 billions flows per second
500 trillion connections!

# Aphasia
# Resource
# Materials

# APHASIA Car Magnet

Aphasia is an acquired language disorder caused by a stroke or a brain injury. Aphasia impairs a person's ability to communicate, but does not affect their intelligence.

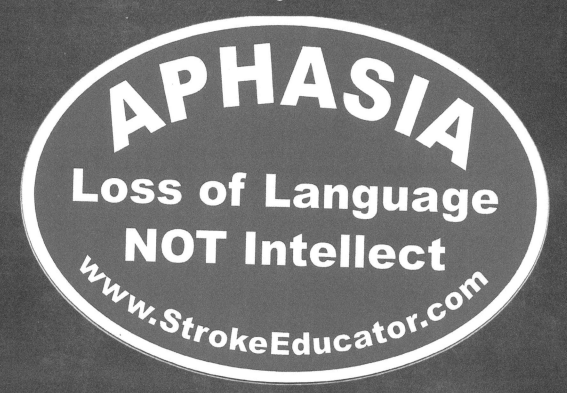

APHASIA
Loss of Language
NOT Intellect
www.StrokeEducator.com

One of the easiest ways to help educate the public about aphasia is with a car magnet. Many organizations (such as Adler Aphasia Center, Stroke Educator, Inc.) buy aphasia magnets in bulk and sell or donate them to their customers.

# APHASIA Research

The *Aphasia in North America* white paper listed "Insufficient <u>awareness and knowledge</u> of aphasia by health care providers and the wider public" as the #1 gap in aphasia awareness (Simmons-Mackie, 2018).

**APHASIA IN NORTH AMERICA**

NINA SIMMONS-MACKIE, PH.D.

FREQUENCY
DEMOGRAPHICS
IMPACT OF APHASIA
COMMUNICATION ACCESS
SERVICES AND SERVICE GAPS

**AphasiaAccess**
Until Every Voice is Understood

Aphasia affects over *two million Americans* and is *more common than Parkinson's disease, cerebral palsy, or muscular dystrophy.*

Nearly 180,000 people acquire the disorder each year. *<u>However, most people have never heard of it</u>* (National Aphasia Association).

# APHASIA ID Cards

It is always helpful to use an aphasia ID card, with information about aphasia, when a person is having problems communicating and can't explain his/her needs (whether ordering a meal or asking for directions).

**APHASIA ID**

Thomas G. Broussard, Ph.D.
Stroke Educator, Inc.

*Person with aphasia*

Date Issued:
11/30/2019

Aphasia is an impairment of language, not of intellect. Aphasia can affect a person's speech and their ability to read or write and is usually the result of a stroke or another brain injury. People with aphasia have trouble communicating.

The following tips help:

➔ Please be patient with me.

➔ Please speak slowly and in a normal voice. No need to shout!

➔ Try to avoid loud places or places with a lot of background noise.

Thank you very much for your patience and understanding.

Visit aphasia.org for more information and aphasia-friendly resources.

Made at www.aphasiaID.com

**I had a Stroke—Aphasia**

I have trouble with numerals and words
My name is Tom Broussard
My address is ...
Waltham, MA 02452
Laura Broussard is my wife # ...

Get a free aphasia ID card from any number of organizations, including Lingraphica, the National Aphasia Association, or you can make one for yourself.

# APHASIA Caregiver Resources

Lingraphica, the National Aphasia Association, and other organizations provide excellent resources for the challenges faced by family and caregivers of people with aphasia.

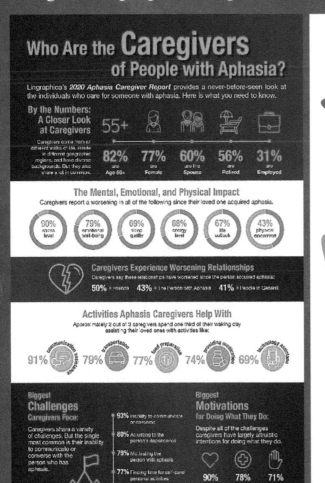

National Aphasia Association

## the aphasia caregiver guide

ADVICE FOR NAVIGATING APHASIA AND YOUR LOVED ONE'S CARE WITHOUT LOSING YOURSELF ON THE JOURNEY

by the National Aphasia Association

# APHASIA Website Resources

- Adler Aphasia Center https://adleraphasiacenter.org/

- American Heart Association (AHA) / American Stroke Association (ASA) https://www.heart.org/

- American Speech-Language-Hearing Association (ASHA) https://www.asha.org/

- Aphasia Access https://www.aphasiaaccess.org/

- Aphasia Center of California http://www.aphasiacenter.net/

- Aphasia Center of Maine https://www.aphasiacenterofmaine.org/

- Aphasia Institute https://www.aphasia.ca/

- Aphasia Recovery Connection (ARC) https://www.aphasiarecoveryconnection.net/

- Aphasia Tool Box https://www.aphasiatoolbox.com/

- Boston University Aphasia Resource Center http://www.bu.edu/aphasiacenter

- Brooks Rehabilitation Aphasia Center (BRAC)
  https://brooksrehab.org/locations/aphasia-center/

- Lingraphica https://www.aphasia.com/

- National Aphasia Association (NAA) https://www.aphasia.org/

- Tactus Therapy https://tactustherapy.com/

- The League for People with Disabilities, Inc.
  https://www.leagueforpeople.org/scale

- The Shirley Ryan AbilityLab https://www.sralab.org/

- Teaching of Talking https://teachingoftalking.com/

- Triangle Aphasia Project (TAP) Unlimited
  https://www.aphasiaproject.org/

- Stroke Comeback Center https://strokecomebackcenter.org/

- Stroke Educator, Inc. www.strokeeducator.com

- UCF Aphasia House
  https://healthprofessions.ucf.edu/cdclinic/aphasia/

- Voices of Hope for Aphasia http://www.vohaphasia.org/

# APHASIA Recovery Plan Tips

**Start the Plan as Soon as Possible** – Start before conventional speech therapy ends, using all the energy and direction available from a team of speech therapists, family, and caregivers.

**Time Is the Issue** – Recovery takes time and is measured in terms of, among other things, interest, motivation, practice, effort, and awareness. It is a lifelong effort to improve one's language disruption from aphasia.

**Build It and They Will Come** – Build all the language activities, and improvement will come. This includes: joining your local aphasia support groups; writing, reading, talking, and walking daily; recording, listening, and studying your voice; taking pictures of the world around you; and joining an intensive program or aphasia boot camp.

**Mission Impossible: Your Mission, Should You Choose to Accept It** – Stroke and aphasia recovery can feel like the movie when the mission is basically impossible, yet the team does it and wins anyway. Others often think it can't be done, but people who have lost their language still have the skills to do the "impossible" and regain their language.

"Experience is what you get when you don't get what you want!"

—all people with
or without aphasia

# Glossary

**aac (augmentative and alternative communication) devices.** Devices that offer communication methods and machines used to supplement or replace typical communication modes (i.e., writing or speech) for people with aphasia (and other brain diseases).

**active ingredients.** Active ingredients used in drugs are biologically active. When it comes to aphasia, the active ingredients needed for recovery are found in the use of continual, repetitive, and persistent language activities (reading, writing, and speaking), plus exercise that induces neuroplasticity.

**activities.** Therapeutic "activities" for people with aphasia are *themselves* the active ingredients that induce and "turn on" plasticity and the resultant recovery. Speech therapy for each deficit *requires* action to regain lost skills. Therapy for writing *requires* constant writing. Therapy for reading *requires* constant reading, and so on.

**agnosia.** People with agnosia are unable to process sensory information. Often there is a loss of ability to recognize objects, persons, sounds, shapes, or smells while the specific sense is not defective nor is there any significant memory loss. Agnosia causes the victims to lose the ability to recognize or comprehend the meaning of objects, even with intact senses.

**agrammatism.** Characteristic of nonfluent aphasia. Individuals with agrammatism have speech that is often characterized by content words without function words. For example, when asked to drive to the store, a person with aphasia might say, "Store … drive … now," not unlike telegraphic speech, with simple sentences (with many or all function words omitted), akin to telegraph messages.

**agraphia.** An acquired neurological disorder, like stroke and aphasia, causing an inability to communicate through writing, either due to some form of motor problem or an inability to spell.

**alexia.** A rare neurological disorder marked by loss of the ability to understand written or printed language, usually resulting from a brain lesion or a congenital defect.

**amnesia.** A general term that describes memory loss. The loss can be temporary or permanent. The causes can include head and brain injuries, certain drugs, alcohol, traumatic events, or conditions such as Alzheimer's disease.

**anarthria.** A severe form of dysarthria; a motor speech disorder that occurs when someone can't coordinate or control the muscles used for speaking. People with anarthria cannot articulate speech at all. The condition is usually a result of a brain injury or a neurological disorder, such as a stroke or Parkinson's disease.

**anomic aphasia.** A mild, fluent type of aphasia where individuals have word-finding problems and cannot express the words they want to say (particularly, nouns and verbs). It is a deficit of expressive language and usually seen in many people with aphasia.

**aphemia.** A term formerly used to describe a type of motor aphasia and more recently used as a synonym for apraxia of speech.

**aphonia.** The inability to speak due to damage to the nerves and the vocal cords, as a result of surgery or a tumor. People with this disorder have "lost their voice."

**apraxia of speech (AOS).** A developmental or an acquired motor speech disorder (likely, a brain injury or a stroke) affecting an individual's ability to connect speech messages from the brain to spoken words, leading to limited or difficult speech ability.

**art therapy.** An art discipline that incorporates creative methods of expression through various art media. People with aphasia are often unable to express themselves, yet use art (drawings, painting, and sculpture, among other things) as a way to express their interests, ideas, and goals. These activities can contribute towards inducing plasticity and any resultant learning.

**axon.** A nerve fiber, a long, thin "wire" of a nerve cell, or neuron. It conducts electrical impulses away from the body of the nerve cell. Its function is to transmit information to different neurons, muscles, and glands in the brain.

**body language.** A type of nonverbal communication that is often used by people with stroke and aphasia. While still having problems expressing their needs, people use body language to help transmit much of the needed information. It is a subset of nonverbal communication and complements verbal communication in social interactions.

**Broca, Paul.** Pierre Paul Broca (1824–1880) was a French physician, anatomist, and anthropologist. He is best known for his research on the Broca's area, a region of the brain's frontal lobe that is named after him. Broca's area is involved with language and is often referred to as the "language center."

**Broca's aphasia.** A speech and language problem usually resulting from a stroke. People with Broca's aphasia have a problem finding words and have an inability to complete full sentences when speaking or writing. It is exemplified by slow, halting, or interrupted speech, with noticeable pauses when trying to speak. (See **expressive aphasia**, **telegraphic speech.**)

**CAT scan.** CAT scan (computerized axial tomography scan), or CT scan, or computed tomography scan, uses a computer-processed combination of many X-ray measurements taken from different angles to produce cross-sectional images (virtual "slices") of specific areas, to see the inside of an object without surgery.

**cell.** See **neuron.**

**cerebrovascular accident (CVA).** Known as a stroke, it is an injury of the brain where the blood supply to a part of the brain is interrupted, either by a clot in the artery or an artery bursts.

**circumlocution.** A phrase that is often referred to as "going around the barn," or roundabout speech. People with aphasia have a difficult time finding the *one* word they *want* to say, but find lots of *other* words to help describe what they are *trying* to say. For example: A person was pointing and waving at an urn, but couldn't say the word *urn*. He finally blurted out, "I want that thing that is used for that thing for people who are dead and burned."

**conventional therapy.** The beginning of language therapy which includes naming, repetition, sentence completion, following instructions, and conversations. Conventional language therapy (approximately 50 hours) is the start of helping patients relearn communication to lessen deficits (such as in writing, reading, and speaking) that were caused by stroke and aphasia.

**cryptogenic stroke.** In most cases, a stroke is caused by a blood clot that blocks blood flow to the brain. But in some instances, despite testing, the cause can't be determined. Strokes without a known cause are called "cryptogenic."

**dendrites.** Small branched extensions ("short wires") of a nerve cell that deliver electrochemical stimulation and transmit information to other dendrites by upstream neurons. Dendrites play an important role in integrating the inputs before sending the newly integrated output to other downstream receptors.

**drawing.** People with aphasia who are unable to produce written words use drawings, charts, graphs, or metaphorical drawings to convey their needs and thoughts without using the necessary words.

**dysarthria.** A condition in which the muscles used for speech are weak or there is difficulty controlling the muscles. It is often characterized by slurred or slow speech that can be difficult to understand.

**dysgraphia.** A learning disability and a neurological disorder that affects writing abilities. It is described as the difficulty of handwriting, spelling, and thinking and writing at the same time, plus having trouble putting thoughts to paper. The disorder typically causes a person's handwriting to be distorted and difficult to understand.

**dysnomia.** A learning disability and a neurological disorder of remembering names or words from memory that are needed for spoken or written expressive language. The individual may be able to describe an item, but be unable to recall the specific name of the item. It is considered a milder version of anomia. (See **anomia.**)

**dysphagia.** The problem of not being able to swallow solids without coughing or choking.

**enriched environment.** The added stimulation of the brain by its physical and social surroundings. More stimulating environments lead to higher rates of (synaptic) growth (leaves), and more complex branching (dendrite arbors), which lead to increased brain activity, growth, and learning.

**enriched therapy.** Conventional therapy (50 hours) is the start to aphasia recovery; intensive therapy (150 hours) is a sprint; and enriched therapy (1,500 hours per year) is a marathon. These activities include persistent reading and writing, regular social interaction and speaking, plus consistent exercise and walking. The activities provide the lifelong communication (lifelong learning) marathon needed to induce plasticity and the resultant learning.

**evidence.** Keeping track and saving the results of various communication activities allows for reflection, assessment, and feedback of ongoing (and still changing) outcomes. Examples include: keeping and reviewing a diary, listening to your voice recordings, taking and viewing pictures, taking and watching one's video recording, and reviewing one's exercises.

**executive function.** Includes basic brain processes such as attention control, cognitive inhibition, inhibitory control, working memory, and cognitive flexibility. People with aphasia have deficits with higher-order executive functions including planning, reasoning, and problem solving.

**experience-dependent neuroplasticity.** Also known as brain plasticity or neural plasticity. The brain has the ability to change continuously throughout an individual's life. People with stroke and aphasia have the capacity to regain their language (and other functions), based on therapeutic (reading, writing, speaking) activities, plus exercise, all of which induce plasticity.

**feedback.** Occurs when outputs of a system are converted into inputs as part of a cause-and-effect loop of enhanced awareness and learning. People with aphasia need to establish a feedback loop for every language deficit in order to "see what you *wrote*," "hear what you *said*" and "watch what you *saw*." Without gathering and saving the evidence, there would be no instructive lifelong feedback loop, resulting in less learning.

**flashcards.** Flashcards (or flash cards) provide information with text and/or images to help with language deficits such as "automatics" (problems with the sequence of letters, numbers, days of the week, months of the year, and so on) for people with aphasia. This is often the first step for regaining one's language from a stroke—by providing instant recognition and feedback.

**fMRI (functional magnetic resonance imaging).** Measures brain activity by detecting changes in the blood flow coupled with brain activities. It is a brain scan used to map neural activity in the brain or spinal cord, related to energy use by the brain cells.

**gestures.** A form of nonverbal communication with visible bodily actions that communicate particular messages, either in place of, or in conjunction with speech.

After a stroke and aphasia, gestures are often the first step of communicating needs before language can be restored.

**global aphasia.** A severe form of nonfluent aphasia, caused by damage to the left side of the brain, that affects both written and oral language as well as auditory and visual comprehension, with significant impairments across all aspects of language, including impaired speech, comprehension, repetition, naming, reading, and writing.

**habit.** Usually an unconscious set of actions acquired through frequent repetition, such as, "It is a habit never to be late." People with aphasia still have many of the habits they acquired before their stroke. The existing habits, although still unconscious in nature, can be a huge influence on language activities leading to plasticity and recovery.

**high-tech applications.** (See **aac.**)

**ICAPs (intensive comprehensive aphasia programs).** Intensive and comprehensive aphasia programs (aphasia camp or boot camp) that usually provide about 30 hours a week, for 4–6 weeks, of individual and group therapy to people with mild to severe aphasia, to help with language deficits by using repetition and socialization.

**infarct (or infarction).** The tissue death (loss of cells) as a result of inadequate blood supply to an affected area. It is caused by an artery blockage or rupture, and the resulting lesion (see **lesion**) is referred to as an infarct, typically from a stroke.

**information capacity deficit.** An auditory problem such that you cannot receive (hear) and process (understand) messages at the same time. Letters and numbers can appear too fast to be able to process and hear the next letter or number, all within the short time allowed at normal conversational speed. It is a deficit for people with aphasia (and other brain-related problems), but rarely cited.

**intensive therapy.** Provides a certain amount of therapeutic hours (approximately 150 hours) that are considered intensive in a certain period of time. Most intensive therapy programs provide 30 hours per week for 4–6 weeks.

**language deficits.** After stroke and aphasia, and after being told that you have "lost your language" may be the first realization that your language is damaged such that you cannot read, write, or speak well in various proportions. It is also possible that you may be unaware of your deficits.

**lesion** (or **brain lesion**). Describes the damage or destruction to any part of the brain. Trauma, damage, or diseases can cause inflammation or destruction of brain cells, or brain tissue. A lesion can be localized to one place or widespread.

**lexicon.** The vocabulary of a person, language, or branch of knowledge (such as nautical or medical). A sailor's lexicon is centered on ships and the sea. People with aphasia, and their family and caregivers, need to learn a new lexicon about stroke, aphasia, plasticity, and recovery.

**listening.** Along with speaking, reading, and writing, listening is one of the "four skills" of language learning. The act of listening involves complex affective (feeling, attitudes), cognitive, and behavioral processes, so some problems might not have anything to do with one's hearing ability.

**lost cells.** A stroke destroys a large number of brain cells (neurons), ranging from hundreds of millions to billions of cells, and hundreds of miles of myelinated fibers, depending on the severity of the stroke. Recovery comes from the remaining cells that have the capacity to grow the new connections that the lost cells took with them.

**low-tech applications.** Interventions or activities (that are still highly therapeutic), with little or no cost or difficulty, using diaries, journals, voice recordings, video recording, calendars, photography (taking pictures), card games, battery-powered games, computer games, flash cards, and more.

**ministroke.** Also known as a transient ischemic attack (TIA). It occurs when part of the brain experiences a temporary lack of blood flow, with stroke-like symptoms that typically resolve within 24 hours. Ministroke symptoms (or TIAs) and stroke symptoms are nearly identical, so immediate emergency attention is needed. About 1 in 3 people who experiences a ministroke might have a full stroke later, so prevention is the key.

**mixed nonfluent aphasia.** Applies to people who have sparse and effortful speech, resembling severe Broca's aphasia. However, unlike individuals with Broca's aphasia, mixed nonfluent aphasia patients also have very poor comprehension of speech, similar to people with Wernicke's aphasia.

**MRI (magnetic resonance imaging).** A medical imaging technique used in radiology to form pictures of the anatomy and the physiological processes of the body. MRI scanners use strong magnetic fields to generate images of the organs in the body. MRI does not involve X-rays, which distinguishes it from CT and PET scans.

**music therapy.** A type of expressive arts therapy that uses music to improve and maintain the physical, psychological, and social well-being of individuals. When a person experiences difficulty communicating after a stroke, singing words or short phrases set to a simple melody can often enhance speech production and fluency.

**myelin sheath.** A fatty substance that surrounds nerve cell axons (the nervous system's wires) to insulate them and increase the speed of information of electrical impulses passing through the axon. The myelinated fiber can be likened to an electrical wire (the axon) with insulating material (myelin) around it.

**neologism.** Used in a combination of two existing words or used to shorten or distort existing words that only have meaning to the person who uses them. The use of neologisms may also be due to aphasia, after brain damage from a stroke or head injury.

**neurotransmitters.** A chemical messenger that carries, boosts, and balances signals between neurons and other cells in the body. These messengers affect a wide variety of physical and psychological functions and are part of shaping thoughts and actions.

**node of Ranvier.** Known as the gap (or break) in the myelin sheath, it occurs along a myelinated axon where the gap is exposed to the outside space. The gap is uninsulated, and that allows faster conduction of the electrical impulses.

**numeracy.** The ability to reason and to use simple numerical concepts like addition, subtraction, multiplication, and division. People with aphasia may or may not have problems with the numbers themselves. But there still might be problems converting the numbers into the written equivalent of those numbers. Adding up numbers to pay a bill is one thing, but writing the check for the written equivalent of those numbers is another.

**paraphasia.** A type of language problem commonly associated with aphasia, and characterized by the introduction of extra or unintended syllables, substituting one word for another, or nonsensical phrases while the speaker is typically unaware of the errors.

**PET (positron emission tomography) scan.** An imaging test that helps reveal how a person's tissues and organs are functioning. A PET scan uses a radioactive drug (tracer) to show this activity. This scan can sometimes detect disease before it shows up on other imaging tests.

**physical therapy (PT).** Treats conditions such as chronic or acute pain, soft tissue injuries, cartilage damage, arthritis, gait disorders, and physical impairments typical of musculoskeletal, cardiopulmonary, and neurological issues, and is designed to improve a patient's physical functions.

**pictures.** Taking pictures is another therapeutic component that stimulates thinking and learning. People with aphasia who are unable to fully tackle reading, writing, or speaking because of their deficits, can take pictures of the world around them and review the images repetitively, later at home. This provides more cognitive understanding with every single glance.

**plasticity ink™.** A metaphorical device of plasticity and the ability of the brain to change at any age, indicating that all learning, whether healthy or not, is written in plasticity ink™.

**plateau.** Described as a period of activities with little or no improvement or learning. People with aphasia are sometimes described by clinicians as "plateauing," when there is not enough progress or clinical improvement. However, the people *with* aphasia themselves never describe themselves as "plateauing," given that their lifelong day-to-day learning is the basis for improvement, no matter how small.

**practice.** The five rules of aphasia recovery are motivation, and practice, practice, practice, and more practice! The secret to recovery is *more* practice, by doing almost anything of interest that piques your curiosity and drives you to satisfy your need for constant study and reading. The resultant learning then urges even *more* practice and study.

**principles of neuroplasticity.** The 10 principles include:

Use it or lose it.
Use it and improve it.
Specificity.
Repetition matters.
Intensity matters.
Time matters.
Salience matters.
Age matters.
Transference (or generalization of skill or activity).
Interference.
These principles contribute to plasticity and learning.

**prior knowledge.** People use their prior experiences and knowledge to learn something new. People with aphasia still retain their past relevant experiences with which to attack the new problems of the reading, writing, or speaking deficits. Knowing what you know and connecting new knowledge to old is the foundation for subsequent learning.

**reading aloud.** Combines the deficits of reading and speaking. Reading aloud helps with comprehension, memory, and the difficulty and timing of speaking at conversational speed. It also helps you to become more aware of the problems, by studying the spoken deficits as they reverberate in the mind's eye.

**receptive aphasia.** A type of aphasia for people who have difficulty understanding the meaning of written and spoken language. People with this condition also speak fluently, but with words that don't make much sense.

**repetition.** One of the principles of neuroplasticity. But repetition doesn't mean that you have to repeat a word over and over again. Saying a word, writing the word, then seeing a picture of the same item induces plasticity, which builds a neural copy in the brain such that one can remember the word better and faster with more integrated action.

**rise time** (or **slow rise time**). An auditory process deficit for people with aphasia that tends to make them miss the first 2 or 3 words of a message as a result of the slow shift from passive to active listening. One solution to the problem when talking with a person with aphasia is to use an alert, saying, "Hey, Tom," before starting the conversation. The person with aphasia will be prompted to hear what is about to be said, given the head start the alert provides.

**salience.** One of the principles of neuroplasticity. The more noticeable, important, or memorable an item is to a person, the more likely they are to remember the item or activity since it "stands out" among the many.

**severity.** The severity of a stroke is measured by the NIH (National Institutes of Health) Stroke Scale, which assesses several aspects of brain function, including consciousness, vision, sensation, movement, speech, and language. Stroke severity uses a scale scoring system: minor stroke (1–4), moderate stroke (5–15), moderate/severe stroke (15–20), and severe stroke (21–42).

**short-term memory.** The capacity to hold a small amount of information in an active, readily available state for a short period (a few seconds) of time. Reciting a phone number is an example of short-term memory, although remembering a phone number from years ago means it has moved to long-term memory.

**silent strokes.** Small strokes that occur without any of the symptoms typical of recognizable stroke, such as paralysis or speech problems.

**stroke.** A medical condition in which obstructed blood flow to the brain results in cell death. There are two main types of stroke: ischemic (85%), due to lack of blood flow; and hemorrhagic (15%), due to bleeding. Symptoms of a stroke include an inability to move or feel one side of the body, problems understanding or speaking, dizziness, or loss of vision on one side.

**stroke survivors.** All strokes are different, as are all stroke survivors. The impairments of stroke survivors (vision, balance, speech, hearing, and paralysis on one side) have their own assortment of injuries in different degrees. About 25% to 40% of people with stroke acquire aphasia.

**synapses.** Connections that make the brain work. A synapse consists of the outgoing terminal (of the sending cell), the space (gap) between, and the incoming terminal (of the receiving cell). The outgoing terminal converts an electrical impulse to a chemical messenger (neurotransmitter) that floats across the space and transmits the message to the next cell.

**syntax.** The set of rules, principles, and processes that govern the structure of sentences in a given language, including word order. A stroke (for those who have lost their language) has damaged the neurological process that operates the rules and sentence structure of the syntax.

**therapeutic.** Can be either a therapeutic drug, or a therapeutic exercise, or something considered therapeutic that helps heal or restore health. All language activities provided by speech therapists and their patients are considered "therapeutic," as well as practicing reading, writing, and speech activities while working alone or with family and friends.

**tissue plasminogen activator (tPA).** A protein involved in the breakdown of blood clots. It is an enzyme found on the inner lining of blood vessels and the cells that line them. It brings about the conversion of plasminogen to plasmin, the major enzyme responsible for clot breakdown.

**transient ischemic attack (TIA).** A neurological incident caused by a brief loss of blood flow in the brain, spinal cord, or retina, but resulting in little or no tissue death. TIAs have the same underlying issues as ischemic strokes caused by a disruption in cerebral blood flow, with the same symptoms: weakness or numbness on one side of the body, loss of vision, difficulty speaking, slurred speech, and confusion.

**video recording.** Helps people with aphasia by recording their actions. They later observe what they have been doing and saying. It offers a new neurological perspective that provides the feedback to "see" what one has been unable to see (or hear) before. Feedback is a very important therapeutic facet of recovery.

**walking.** An important component of brain health, plasticity, and recovery, given that walking (and various kinds of exercise) induce a number of neurotransmitters that are particularly well suited to assist with the process of plasticity at the cellular level.

**Wernicke, Karl (Carl), (1848–1905).** Was a German physician, anatomist, psychiatrist, and neuropathologist. He is known for his study of the localization of brain function and receptive aphasia commonly associated with Wernicke's area, and Wernicke's aphasia was named after him.

**Wernicke's aphasia.** (See **receptive aphasia.**)

**Wernicke's area.** The area of the brain involved in the receptive (as opposed to expressive) component of comprehension of written and spoken language. Damage caused in the Wernicke's area means that a person with aphasia is able to (fluently) speak individual words and sentences, but with phrases and sentences that have little or no meaning.

**word-finding.** The most common problem for people with aphasia is not being able to express a word they want to say. Instead, people with word-finding problems use empty words to fill in the missing word in a sentence, with: *that, that thing, you know, it.* There are word-finding recall strategies using *who, what, why, where,* and *when* questions, to help describe, write, or draw the item at the center of the hunt. Literally, the more you *search* for that word, the closer you get to the target.

# Aphasia Recovery

"Now is not the end.
It is not even the
beginning of the end.
But it is, perhaps, the
end of the beginning."

—Sir Winston Churchill

## Author's Bio

## Thomas G. Broussard, Jr., Ph.D.

Three-Time Stroke Survivor and *Johnny Appleseed of Aphasia Awareness*

Dr. Broussard was an associate dean at The Heller School at Brandeis University until his stroke in 2011. After the stroke, he could not read, write, or speak well. He started Stroke Educator, Inc., with a national Aphasia Awareness Campaign dedicated to educating the wider public about aphasia, a life-changing language disorder that few people have ever heard of!

Since then, he has spoken (in person and online) in 26 states, giving over 225 presentations to more than 5,000 people with stroke and aphasia, and to their families, caregivers, clinicians, and students.

## Other Books

*Stroke Diary: A Primer for Aphasia Therapy* is practically a day-to-day diary from a stroke survivor who couldn't write—but kept on writing anyway. A first-of-its-kind primer that blazes the trail for new aphasia therapy.

*Stroke Diary: The Secret of Aphasia Recovery* is a personal, intensive, enriched therapy boost for recovery, drawn from an almost 500-page diary. The secret of recovery from aphasia is all about the *doing*.

*Stroke Diary: Just So Stories ... How Aphasia Got Its Language Back* validates how practice can provide the cure to aphasia recovery. Practice is more than just *practice*. Practice is the prescription for improvement *and* the cure.

CPSIA information can be obtained
at www.ICGtesting.com
Printed in the USA
BVHW021336300820
587598BV00003BA/4